MIRRORS

HOWARD McCORD | CHERYL DOERING
poems | drawings

We thank the publishers who first
printed some of these poems,
including *Granite* and *Lillabulero*.

Published by James Bertolino
in an edition of 700 copies
at Stone-Marrow Press, Ithaca, New York

Distributed nationally by
Serendipity Books, 1790 Shattuck Ave.,
Berkeley, California 94709

Please direct orders to them.

MIRRORS

RATIOS

I would live in a lightning bolt,
a man-salmon thick in roiling,
dizzy light, a cosmic surf,

its rush like water's
an instantaneous life
gone through

as my own gasp breaks.

Before breath comes
I have been,
was

        that universe
dissolved   but
time eternal there
as anywhere.

LATRODECTUS MACTANS

The water-meter housing
in the front yard
is full of black widows.
If you put your head
inside you can listen
to their spider-whispers.
They sing in voices
sweet as corn
and will spin coverlets
for your eyelids.
If you put your hand inside
they will snap and bite
your fingertips
and try to feed you to
their babies.

Their language is complex,
like tangled knots
of cedar roots,
and takes many years
to learn.

MY SON'S SPEECH

At Tūs it was said that certain
pious children, on first pronouncing
a word, might unleash its primordial
energy and manifest the power of the
Λογος.

"Asherah, asherim," my son
chants, "I will that runes
be riven into wood."
Each word is an angel
of immense authority.
My son knows words
that make ice
when said for the first time,
others that will set a fire
in stone.

He seldom speaks,
but broods over the dictionary
and dreams of the Ice Age
he will create with his tongue,
and of the smoke that will curl
from rock and water.
At his small desk the universe
is regarded affectionately,
carelessly, like a chick
in the palm.
My son has said he will call
the Moon one day
and she will come.

MARCH

What can we do
about the absence
of flowers?

Hoarfrost's
a blossom

but smells
of ice.

All hinges break,
the clock

wont last.

OUTLINE OF THE CURSE NOW IN EFFECT

A slave in a silver mine
on the moon, a prisoner
of the heart's requirement
to beat and break,
a hungry man given no food.

Time is a hammer,
a noose, yet you
still call for love.

Fool, jackanape,
a toad-warp on you,
may the alphabet fall
from your mouth
and the stones of longing
grow like tumors
from your bones.

You can be cleaned to
await the next one.

EXORCISM

It's not that I'm afire,
but that shadows beat
through me like geese
flying high in the night.
They come from so very far
away I cannot say its
name. It may have none.

I am possessed, and know
what possession means.

Nothing possesses me.

Somehow I clench my own
mind in my fists,
but what is there
I may not touch
or know.

Your name, spirit,
is ________

What goes in the blank?

I will kiss the stone
you hold in your hand,
take all that cold inside...

If you will tell
me the first three
letters
of your name.

FOR THEM

The moon is wet.
The stars are
all of water.

Only the wind
through my heart
is dry.

PAST DUSK

A bird sleeps
alone on a branch.

There are no
leaves.

Wild birds
seem not
to sleep
together.

Venus rides
out
beyond us
all

reflecting light,

setting my
heart
just on
yours

2000 miles
away.

THE LEAST DEMAND

There are mountain ranges
that rise within
the eyes, cut off
horizons (make them)
as they offer canyons.

Nobody knows how
they grow.
Beautiful, fascinating
cancers of uncertainty.

They may not really
be there.
But you must find
the Pass
to make the other
side.

ALEXANDRIA

"Order," she said,
lips like scissors,
"Order, a direct
route."

A one-way map
to the frontier.
The map ticks like a watch.
It blooms in spring,
falsely.
The bottom of the sea
is made of wire
in coils.

In the city, I look
down through the streets
a hundred feet,
pipes and tunnels
swimming everywhere.

We are going together.

Glad Day Press
UNION SHOP I.U. 450
308 Stewart Ave., Ithaca, N.Y. 14850